ACADEMIC PLANNER
2022-2023

YEARLY GOALS 2022-2023

AUGUST 2022	**FEBRUARY 2023**
SEPTEMBER 2022	**MARCH 2023**
OCTOBER 2022	**APRIL 2023**
NOVEMBER 2022	**MAY 2023**
DECEMBER 2022	**JUNE 2023**
JANUARY 2023	**JULY 2023**

IMPORTANT DATES

WEEK TIMETABLE

TIME	S	M	T	W	T	F	S

ASSIGNMENT LOG

ASSIGNMENTS	CLASS	DUE DATE	GRADE

ASSIGNMENT LOG

ASSIGNMENTS	CLASS	DUE DATE	GRADE

ASSIGNMENT LOG

ASSIGNMENTS	CLASS	DUE DATE	GRADE

MY NOTES

MY NOTES

AUGUST 2022

IMPORTANT THIS MONTH

GOALS

DATES TO REMEMBER

MY NOTES

AUGUST 2022

SUNDAY	MONDAY	TUESDAY	WEDNESDAY
	1	2	3
7	8	9	10
14	15	16	17
21	22	23	24
28	29	30	31

NOTES:

THURSDAY	FRIDAY	SATURDAY	TO DO
4	5	6	______________
11	12	13	______________
18	19	20	______________
25	26	27	______________
			______________

NOTES:

____

____

SEPTEMBER 2022

IMPORTANT THIS MONTH

GOALS

DATES TO REMEMBER

MY NOTES

SEPTEMBER 2022

SUNDAY	MONDAY	TUESDAY	WEDNESDAY
4	5	6	7
11	12	13	14
18	19	20	21
25	26	27	28

NOTES:

SEPTEMBER 2022

THURSDAY	FRIDAY	SATURDAY	TO DO
1	2	3	
8	9	10	
15	16	17	
22	23	24	
29	30		

NOTES:

OCTOBER 2022

IMPORTANT THIS MONTH

GOALS

DATES TO REMEMBER

MY NOTES

OCTOBER 2022

SUNDAY	MONDAY	TUESDAY	WEDNESDAY
2	3	4	5
9	10	11	12
16	17	18	19
23	24	25	26
30	31		

OCTOBER 2022

THURSDAY	FRIDAY	SATURDAY	TO DO
		1	____________
6	7	8	____________
13	14	15	____________
20	21	22	____________
27	28	29	____________

NOTES:
____

____

NOVEMBER 2022

IMPORTANT THIS MONTH

GOALS

DATES TO REMEMBER

MY NOTES

NOVEMBER 2022

SUNDAY	MONDAY	TUESDAY	WEDNESDAY
		1	2
6	7	8	8
13	14	15	16
20	21	22	23
27	28	29	30

NOTES:

NOVEMBER 2022

THURSDAY	FRIDAY	SATURDAY	TO DO
3	4	5	
10	11	12	
17	18	19	
24	25	26	

NOTES:

DECEMBER 2022

IMPORTANT THIS MONTH

____

____

____

____

GOALS

____

____

DATES TO REMEMBER

____

____

____

MY NOTES

DECEMBER 2022

SUNDAY	MONDAY	TUESDAY	WEDNESDAY
4	5	6	7
11	12	13	14
18	19	20	21
25	26	27	28

NOTES:

DECEMBER 2022

THURSDAY	FRIDAY	SATURDAY	TO DO
1	2	3	
8	9	10	
15	16	17	
22	23	24	
29	30	31	

NOTES:

JANUARY 2023

IMPORTANT THIS MONTH

GOALS

DATES TO REMEMBER

MY NOTES

JANUARY 2023

SUNDAY	MONDAY	TUESDAY	WEDNESDAY
1	2	3	4
8	9	10	11
15	16	17	18
22	23	24	25
29	30	31	

NOTES:

____

____

JANUARY 2023

THURSDAY	FRIDAY	SATURDAY	TO DO
5	6	7	__________
12	13	14	__________
19	20	21	__________
26	27	28	__________
			__________

NOTES:
____

____

FEBRUARY 2023

IMPORTANT THIS MONTH

GOALS

DATES TO REMEMBER

MY NOTES

FEBRUARY 2023

SUNDAY	MONDAY	TUESDAY	WEDNESDAY
			1
5	6	7	8
12	13	14	15
19	20	21	22
26	27	28	

NOTES:

FEBRUARY 2023

THURSDAY	FRIDAY	SATURDAY	TO DO
2	3	4	
9	10	11	
16	17	18	
23	24	25	

NOTES:

MARCH 2023

IMPORTANT THIS MONTH

GOALS

DATES TO REMEMBER

MY NOTES

MARCH 2023

SUNDAY	MONDAY	TUESDAY	WEDNESDAY
			1
5	6	7	8
12	13	14	15
19	20	21	22
26	27	28	29

NOTES:

MARCH 2023

THURSDAY	FRIDAY	SATURDAY	TO DO
2	3	4	
9	10	11	
16	17	18	
23	24	25	
30	31		

NOTES:

APRIL 2023

IMPORTANT THIS MONTH

____

____

____

____

GOALS

____

____

DATES TO REMEMBER

____

____

____

MY NOTES

APRIL 2023

SUNDAY	MONDAY	TUESDAY	WEDNESDAY
2	3	4	5
9	10	11	12
16	17	18	19
23	24	25	26
30			

APRIL 2023

THURSDAY	FRIDAY	SATURDAY	TO DO
		1	————
6	7	8	————
13	14	15	————
20	21	22	————
27	28	29	————

NOTES:

MAY 2023

IMPORTANT THIS MONTH

GOALS

DATES TO REMEMBER

MY NOTES

MAY 2023

SUNDAY	MONDAY	TUESDAY	WEDNESDAY
	1	2	3
7	8	9	10
14	15	16	17
21	22	23	24
28	29	30	31

NOTES:

MAY 2023

THURSDAY	FRIDAY	SATURDAY	TO DO
4	5	6	
11	12	13	
18	19	20	
25	26	27	

NOTES:

JUNE 2023

IMPORTANT THIS MONTH

GOALS

DATES TO REMEMBER

MY NOTES

JUNE 2023

SUNDAY	MONDAY	TUESDAY	WEDNESDAY
4	5	6	7
11	12	13	14
18	19	20	21
25	26	27	28

NOTES:

JUNE 2023

THURSDAY	FRIDAY	SATURDAY	TO DO
1	2	3	
8	9	10	
15	16	17	
22	23	24	
29	30		

NOTES:

JULY 2023

IMPORTANT THIS MONTH

GOALS

DATES TO REMEMBER

 MY NOTES

JULY 2023

SUNDAY	MONDAY	TUESDAY	WEDNESDAY
2	3	4	5
9	10	11	12
16	17	18	19
23	24	25	26
30	31		

JULY 2023

THURSDAY	FRIDAY	SATURDAY	TO DO
		1	
6	7	8	
13	14	15	
20	21	22	
27	28	29	

NOTES:

CONTACT PAGE

NAME	NUMBER

CONTACT PAGE

NAME	NUMBER

CONTACT PAGE

NAME	NUMBER
_______________	_______________
_______________	_______________
_______________	_______________
_______________	_______________
_______________	_______________
_______________	_______________
_______________	_______________
_______________	_______________
_______________	_______________
_______________	_______________

CONTACT PAGE

NAME	NUMBER

CONTACT PAGE

NAME	NUMBER

PASSWORD PAGE

WEBSITE	PASSWORD

PASSWORD PAGE

WEBSITE	PASSWORD

PASSWORD PAGE

WEBSITE	PASSWORD

PASSWORD PAGE

WEBSITE

PASSWORD

PASSWORD PAGE

WEBSITE   **PASSWORD**

MY NOTES

MY NOTES

MY NOTES

MY NOTES

MY NOTES

MY NOTES

MY NOTES

MY NOTES

MY NOTES

MY NOTES

MY NOTES

MY NOTES

MY NOTES